The Really, Really, Really
Useful Guide

Number 11

VIKING BAY

A NATURAL HISTORY

MIKE PEARCE

MIKE PEARCE

Copyright 2017 by Mike Pearce

All rights reserved. No part of this book may be reproduced, distributed or transmitted in any form or by any means, including photocopying, recording, or other electronic or mechanical methods, without the prior written permission of the author, except in the case of brief quotations embodied in reviews and certain other non-commercial uses permitted by copyright law. You must not circulate this book in any format.

This book may not be resold or given away to other people. Please respect the work of the author and purchase a copy for your own use.

To see other publications by the author visit
snappysnappybooks.com

Copyright © 2017 Mike Pearce

All rights reserved.

ISBN:10:1978433344
ISBN-13:978-1978433342

DEDICATION

This book is dedicated to all visitors to Viking Bay in Broadstairs to help them identify some of the more common plants and animals that live there.

CONTENTS

Acknowledgments

1	Introduction	1
2	Tides	3
3	Chalk	5
4	Flint	7
5	Birds	9
6	Marine life	13
7	Seaweeds (Greens and browns)	22
8	Seaweeds (Reds)	25
9	Flotsam and Jetsam	27
10	Cliff plants	30

ACKNOWLEDGMENTS

The author would like to thank Christine Pearce for reading and checking the manuscript

1 INTRODUCTION

Apart from being one of the best bays in Thanet, Viking Bay in Broadstairs, Kent provides a wide variety of sea shore habitats. The central beach is fine sand, possibly originating from the Goodwin Sands off Ramsgate which needs to be preserved. Looking out to sea from the promenade. One can see to the left near the harbour are scattered rocks of flint often close to the harbour walls. The ridged harbour wall forms a habitat for weeds and shell fish. To the right surrounding a raised paddling pool and extending outwards into the sea is a platform of chalk. Natural pools are found in this chalk which can be wide and shallow or narrow and deep. These form gulleys through which the water from the paddling pool runs back into the sea. Pools and gulleys provide microhabitats allowing animals and plants to isolate themselves away from full exposure if they were on the rock surface. Behind the beach, above the beach huts are exposed chalk cliffs which are colonized by several maritime plants. Each inch of the chalk cliff is said to represent 2,500 years, so walking up the steps from the beach one is passing through a vast period of geological history.

This booklet forms a quick guide to many of the more common animals and plants that can be found in the bay. The amount and kinds of sea shore life present will vary throughout the season, from day to day, from tide to tide and year to year but whatever the time of year one cannot fail to find something new and interesting.

2 TIDES

Two high and two low water periods occur approximately every twenty-four hours (six hours from high to low and six from low to high). The area in the middle of the bay is covered by almost every tide of the year so any plant or animal is exposed to the elements and wave action. High barriers of sand are normally constructed to protect the beach huts during the winter months.

At the lowest shore level is the laminarian zone and the highest the splash zone. Often with an incoming tide the second shore level is covered in one hour and the third covered in three hours (the reverse is true for an outgoing tide). The lowest and highest spring tides occur when the moon is full or new. The best of these is when the sun, earth and moon are in exact alignment at the spring and autumn equinoxes around March and September. Predicted dates for these can be found in Whitaker's Almanac, tide tables or on the internet.

Fresh water also runs down the beach towards the sea. One source is likely from spring water, a spring being present previously at the site of St Mary's chapel in Albion Street.

VIKING BAY (A NATURAL HISTORY)

3 CHALK

On the beach, chalk is eroded by the sea and along the cliffs by rain. Chalk therefore does not form a very secure surface for the attachment of animals and plants especially when large waves force trapped air against the cliffs. Evidence of its removal can be seen in the milky colour of some pools as the tide recedes. Also, the milky, murkiness of the sea water around this coast discourages divers who prefer the cleaner waters in other parts of Britain. Broken chalk pieces can be found on the beach close to the raised chalk. In gulleys, pebbles, especially of chalk are sorted into sizes by weight. Gulleys often occur where faults exist in chalk and are also formed by run off for example padding pools. In some places, small pools are found which are only eleven centimetres wide but can be as deep as one metre. For seaweeds, the mineral content of chalk is not important as chalk is only used as a foothold. Organic acids of shellfish will dissolve chalk. Limpets can go as deep as two and a half centimetres. Worms and borers can burrow or

MIKE PEARCE

travel through small pieces of chalk which also help to break it up.

4 FLINT

Often embedded within the chalk are pieces of flint. They can be various sizes and various shapes. In some cases they can be large and cylindrical or with many protrusions. Flint is sedimentary silica from sponges and plankton deposited over 95 million years ago. It is opaque, glassy and may be covered in a thin layer of chalk. Pieces of flint are dislodged as the chalk is eroded and are found in piles at the base of cliffs or in gulleys on the beach. Flint provides a firm anchorage for sea weeds and marine life. In the gulleys waves can throw flint against the chalk walls wearing them away. Bands of flint can be seen in the cliff as single large pieces or in some cases as flat plates. Fossils from the late Cretaceous period (around 80 million years old) can be found:

Belemnites originate from squid like animals. They are long finger shaped and tapered at one end.

Bivalve pieces are cream coloured flattened plates with regular ridges on the upper surface. These are more common towards Ramsgate harbour.

Conulus originate from sea urchins. They are cone shaped, often filled with flint. One can see lines of indentations running down from the top to the base with very small perforations travelling down these in rows.

Micrasters also originate from sea urchins but these are heart shaped with five star shapes on the upper surface with lines of perforations running from top to bottom.

If some pieces of flint are broken open it is possible to find fossilized sponges inside. Some fossils may be mainly chalk. To preserve these, they must be repeatedly washed in fresh water to remove the salt or they will eventually disintegrate. Areas of flint, some with weed attached, can be found around the harbour and also on one side of the paddling pool. Flint is also embedded in the raised chalk on the beach.

5 BIRDS

Black - headed gulls have black backs and wings and pink legs. They are commonly found at the edge of the water near the chalk. They often feed on marine life which floats to the surface when the tide refloats the seaweed.

Cormorants are large slender black erect birds. They have a long beak and white patches either side of the face. One may see them on the sea and diving below to search for fish or sitting on posts out at sea spreading out their wings.

Curlews have long down curved beaks and brown streak on their feathers. They are often winter visitors to the bay.

Dunlins are small birds with black slightly down curved beaks and wade at the edge of the shore.in winter have a dull greyish upper surface with white underneath. In summer they are reddish brown with a large dark belly patch underneath.

Fulmars are stocky birds, pale grey on the upper side and white below. They have black tips to their beaks and black nostrils (large nasal tubes) on top of their bills. Their eyes are dark. If upset they can shoot out a salt solution at you from their nasal salt glands. They are sometimes found nesting on the rocks near the harbour. In flight they hold their wings stiffly.

Great black gulls are much larger than the herring gull and mainly winter visitors. They often may seek refuge in the bay during stormy weather.

Herring gulls have blue grey backs and wings and pink legs. Many yellow bills have a red spot underneath to encourage young to peck at this spot so the adults regurgitate food. Herring gulls follow the tide, catching crabs and other invertebrates. They feed on dead fish and crabs near the harbour and the strand line and flocks follow fishing boats along the bay. In summer, they may also be seen tapping the ground with their feet in parks to encourage worms to come to the surface. The kew calling note is common as is the trumpeting call made with head raised. Also, the Bub, Bub, Bub call.

VIKING BAY (A NATURAL HISTORY)

In February they pair off around nesting time and often mew like cats. When the young are born, the parents become very protective and when approached give an angry HA, Ha, Ha and fly close to the intruder. From august gulls tend to frequent the beach more rather than on the houses. They pick up scraps from tourists and eat any sea life which is around. In summer you can see them gliding across the bay in the thermals and wind currents. Younger birds can be recognized by their feathers not being pure white.

Pied Wagtails are small sparrow size birds recognisable by their tails moving quickly up and down as they walk. They have black feathers on top of their heads, white patches either side of their heads and white undersides.

Pigeons, house sparrows and crows are often on the beach searching for food.

Sanderlings are small plump birds with black bills and black legs. They can be seen especially in winter scurrying in flocks along the edge of the tide. In winter they are pale grey above with a white head and underside. In summer their head, breast and back can be reddish brown.

6 MARINE LIFE

A lot of deeper sand dwelling organisms are absent because of the underlying chalk preventing burrowing.

Razor shell fish have are long internally grooved thin straight sided shells coloured yellowish brown with a black tinge at the edges. Small circular depressions in the sand near the paddling pool can indicate their presence. Sand has to be deep for their burrowing so numbers in the bay are restricted.

Edible periwinkles like all periwinkles are rolled into a spiral inside. They are stunted, sharply pointed, greyish brown dark with spiral bands. They can be found on chalk and flint especially on raised chalk mainly from mid-shore to deeper water. They have been reduced in the past by early morning collectors...Rough periwinkles are often yellow have ridges on the whorls and found on the upper parts of the shore. Flat periwinkles have a flattened spire no point, a smooth shell and the aperture looks like a tear drop. They can be red green, orange or yellow and often in the middle shore under weed.

Common whelks are one of the largest molluscs and have a dull white shell. They are not very common and are found as empty shells lying on the strandline. They are up to 10cm in length with an oval aperture. They have a wide groove in the front of the shell for the siphon.

Dog whelks are the size of winkles but more elongated and pointed with spiral ridges. They can be many colors: white/yellow/orange/pink/black or striped. They have a thick shell and a groove in front for the siphon. They are carnivorous feeding on winkles, limpets and barnacles.

Whelk egg capsules are yellow, maggot size with pointed tips. They are common in the gulleys often in crevices.

Limpets are volcano shaped with ribs running top to bottom. They are stuck to the rock by their muscular feet so are hard to remove. They are often in depressions they continue to make in the chalk where they return during their life. As with winkles they are common in the chalk gulleys near the paddling pool. Also they are found on the harbour walls often in rows under the concrete ledges and the steps of the promenade but fewer. Larger ones tend to be higher up and on more exposed areas. Odd ones found on pieces of flint are scattered on the shore and harbour.

Thick top shells are colourful pointed spinning top shaped shells with smooth sides. They may have red streaks on a light yellow background. They are sometimes found under stones.

Piddocks are long narrow shells found in holes in chalk which they have made by rotating and rocking their rows of teeth on one end of the shell. The burrow is larger inside the rock to compensate for growth.

Mussels are short, wedge shaped, dark bluish purple, shiny shells tied to rocks. Near the pointed end are numerous strong, hair like threads which attach it to the rocks. Numbers can be few. Very small ones are found in chalk gulleys. There used to be large beds on a lower chalk area to the right of the paddling pool but these no longer exist. They are often completely covered with barnacles even when very young.

Acorn Barnacles have very small whitish pyramid shaped shells with diamond shaped plates opening at the top. These are extremely common on flint, concrete surfaces and the surfaces of shell fish and crabs. Often they are found on the steps of the promenade especially where the horizontal surfaces join the vertical. They tend to be larger further down the shore where they are submerged for longer. There are different species at different zones of the shore.

Edible crabs have a deep, reddish brown, pie shaped shell and black claws. Small ones are found under stones and are very common in gulleys along with the shore crabs where they can be fished for using mussels as bait.

Shore crabs are dark green in colour with markings on the shell but have several white patches when young. They are common under stones and in small pools in gulleys.

Spider crabs have long legs curved forwards and are greyish red. The shell is pear shaped and covered in spines and bristles. Sometimes they are found under stones and empty shells are scattered amongst seaweed in the strandline.

Common prawns are semitransparent with greyish darker tops and are common in pools in late summer. They shoot rapidly across pools if disturbed.

Sand hoppers are grey green in colour with flattened curved bodies and often found under decaying seaweed. They have one long antenna and black eyes. They jump when exposed and burrow into the sand. They are found under stones or weed.

Sea slaters or marine woodlice have flattened, greenish brown, oval bodies with long, bent antennae and are present under stones and amongst the weeds in the strandline.

Sea hares are not common. They look like slugs, snails without shells, and have rabbit like tentacles in front. Their bodies may be colourful and can be found under weed in deeper water on lower parts of the shore.

Sea lemons have a sea slug appearance but have a warty mottled skin with black, red or green patches. They are creamy yellow/brown. They have two tentacles in the front and a rosette of gills at the back. They are sometimes found under stones again in deeper water. There is also a grey variety with lots of protuberances on its back.

Chitons are oval shaped and covered in eight overlapping hard plates surrounded by a ring or belt around the base. They curl up like a woodlouse when detached. They are found under stones sometimes near the harbour wall and cling tightly to rocks.

Common starfish have a red, orange brown, rough upper surface and white area with tubular feet and central mouth below. The tips of the five pointed arms will curl upwards if disturbed. They are common in summer around the harbour walls and in pools in chalk channels near the lower part of the shore.

Small brittle stars have a circular central disc from which five long, slender tubeless, feathery legs radiate.

They are found in muddier sand under stones, especially around the harbour.

Purse sponges appear as little flattened bags open at one end and attached to weeds. They can be found on the sides of the lifeboat slip and collapse to a white flat purse if the inside water is lost

Beadlet anemones are dark red but can be green. They have a jelly like body and short tentacles and are common under flint stones in all regions, especially in gulleys. Tentacles retreat if touched.

Sand mason worms live in flexible transparent paper like tubes covered in sand and shell particles. These protrude above the surface of the sand. The ends are frayed like tentacles. They are found in isolated patches and are seen where sand has accumulated in cavities in chalk and in the sand surrounding pieces of flint. They are common near the harbour.

Sabellaria worms (Sand castle worms) form a mass of raised honeycombs colonies of hard tubes producing solid raised reefs. They are common towards deeper water.

Lug worms are like earthworms with black heads and red bodies. The middle part has bristles. Their presence is seen by spaghetti like coiled worm casts on the sand surface. The other end of its burrow has a small circular depression with a central hole in the sand. Seen in sandy areas on the lower shore.

Pomatoceros keel worms have hard curved white tubes with a ridge on top. They are common on and under flint especially, sometimes covering a large area.

Polydora worms have U shaped burrows recognized as two adjacent very small holes in chalk. They are very widespread in much of the exposed chalk especially at the right of the paddling pool

Blennies are small scaleless fish with large heads and large dorsal fins. Pectoral fins are large and support fish at the front. They are found in paddling pool and deeper pools under weed.

Butterfish are small and flattened sideways. They are light brown with black spots along the top with a long dorsal fin. They are sometimes found under stones near the harbour.

VIKING BAY (A NATURAL HISTORY)

Sprats and other fish are often trapped in the paddling pool as the tide goes out. Larger fish often caught during fishing or washed up dead near the harbour include skate, dogfish, plaice and red gurnard. Often only parts remain after the seagulls have fed on them.

7 SEAWEEDS (GREENS AND BROWNS)

Cladophora has fine yellow/green/brownish tufts/strands like hair. It grows over large areas like a green carpet and is found on the sides of concrete in shade and can reach as high as three metres up the sides of the harbour wall and steps of the promenade. It is also spread over chalk on the beach.

Ectocarpus is a brown unbranched very fine filamentous weed. It is common in pools on chalk.

Spongomorpha has fine branching, tufted, light green strands. It is found on the sides of gulleys and on raised chalk.

Codium tomentosum has thick twisted green strands. It is finger like and well branched and found on chalk.

Sea lettuce looks like bright green wet lettuce leaves with dark veins when older. It is very common on the sides of gulleys. It can reach one foot in length and is very abundant in summer when there is plenty of light.

Serrated wrack has flattened toothed edges with a strong mid rib. It is very common and washed up in large amounts on the beach. It also covers chalk on the raised areas and inner sides of the harbour. Here it forms a slippery surface. The seaweed is also found on pieces of flint on the shore.

Bladder wrack is like serrated wrack but has pairs of air filled bladders sacks along its length to keep it upright in the sea so it gets maximum light. Found on chalk in the middle shore but less abundant.

Channelled wrack may also be found on chalk but pointed bladders are at the tips of the fronds. It is tolerant to exposure and often found at the slash zone.

Bootlace seaweed is a very long lace like weed. It is often slimy and found in strands present in pools in summer and often washed up on shore.

Oarweed laminarian has flat ribbon like hand like

segments on a long hard rubbery stalk. This stalk is attached to an often knobbly thick base. The weed is golden brown often with five to fourteen ribbons. It is found in deep water at the ends of the raised chalk, the ribbon-like fingers being tossed side to side in the currents.

VIKING BAY (A NATURAL HISTORY)

8 SEAWEEDS (REDS)

Polysiphonia is a fine unbranched filamentous red weed found on chalk and edge of pools.

Laurencia pinnatifida is a short stunted red or greenish yellow weed with a swollen stem with small branched shoots either side the branching becoming shorter near the end. Large amounts are found at the top of gulleys and also found on the harbour and in pools.

Dumontia has long thin flattened red strands, often twisted and tapered from the disc like base. It is common in pools.

Geldium is a dark red flattened seaweed, swollen and wider near the tips. It has single side buds. Common in summer on chalk and pools.

Palmaria palmate is purplish red with no side buds. It has flattened thin blades divided into segments. Found on chalk in pools.

Chondus crispus is red and purplish and looks like moss. It has short branches and tufts like a fan with lot of branching. Common on exposed chalk.

Lomentaria articulata is very bright crimson red with flattened chains which look like swollen elongated beads. Found in pools on chalk.

Corralina officinalis is pink with a tufted appearance. It is hardened calcified so the stiff stem supports the branches. Common on the sides of pools. It turns white when dry.

Lithothamnion.is a purple crusty covering found on rocks and stones.

9 FLOTSAM AND JETSAM

This is deposited at the strand line as the tide recedes and around the edges of chalk. Most of the shells of shell fish, invertebrates and seaweeds previously mentioned can be found here especially in winter. Large dead whelk shells are often washed up in the harbour.

Empty egg cases of whelks, seen as hollow cells attached in a ball, are common throughout the year. In the winter these can reach the size of footballs and they may often have other marine life attached such as spiny cockles.

A few examples of things left by the tides not mentioned previously include:

Antennulatia - light brown with hairy brush like hairs sticking up.

Baltic tellins - white, flattish pink or yellow and triangular in shape with thin shells

Breadcrumb sponges - smooth with holes. Yellow and greenish in colour.

Compass jellyfish - transparent with brown and white markings.

Cowries - very small egg shaped shells with flat undersides; black marks on top with long serrated openings.

Cockles - with deeply ridged shells. Shell curving down at hinge.

Cuttle fish bone - white curved at one end with ringed layers underneath.

Dead man's fingers – orange/ yellow fingers, a soft coral.

Dogfish purses - brown elongated long pairs of horns with curly tendrils.

Green sea urchin - spines on shells, central hole in base.

Moon jellyfish - transparent except for four purple rings.

VIKING BAY (A NATURAL HISTORY)

Sea belt - yellow brown, long single pleated ribbon.

Sea gooseberry - transparent with two long tentacles shaped like a gooseberry.

Sea oak - fronds of weed shaped like oak leaves.

Ship worm - wood attacked by boring into wood. Long tunnels with a small hole at entrance.

Skate egg cases - larger than dogfish black and square shaped with short horns either end.

Amongst the weed in summer are often flies. Large black and hairy ones may be Coelopa which can becomes so abundant in early May that they may be also be found in the town. Very small flies can be midges Thassomyi and Ucellia often found on weed. On dead crabs one may find wasps feeding on the meat.

10 CLIFF PLANTS

Two kinds of snail are common on the chalk cliffs on plants. The common garden snail Helix aspera and a narrower smaller white banded snail Helicella.

The most common plants present are members of the cabbage family and grasses. Some examples of plants are as follows:

Sea bindweed is climber/ trailer with shiny leaves and pink trumpet like flowers.

Sea campion has pink and white pot like flower bases and white daisy like flowers.

Wild cabbage has cabbage like leaves with yellow small flowers in clumps. As with many cabbage like plants they are attractive to butterflies especially cabbage white whose caterpillars cause serious damage. Also both snails mentioned will eat the leaves.

Wall rocket has cabbage like leaves with yellow four

petalled flowers on long stems. Seed pods are long and thin, pointing upwards.

Sea beet have green spear shaped or oval glossy. These form a low lying bunched rosette. Flowers are spikes of green. Young leaves or stems have purple patches.

Wild celery has white curved thick green stems with ridges.

Wall flowers are yellow or red flowers, four petalled on erect stalks.

Sea thrift- has a matt of grass like leaves with long stalks ending in a tuft of pink flowers.

Tree mallow /purple lavateria is a shrub with hairy stems and light pink to purple flowers. The flowers have a dark purple centre with radiating purple stripes on the petals. The leaves are sycamore shaped and lower leaves often droop or become brown.

Orache has green, trowel shaped like leaves, the base sometimes reddish brown. The plant is tall with spikes of flowers. The whole plant turns an attractive chestnut brown when dead.

Senecio has yellow flowers, its leaves are silvery greyish and hairy.

Rice grass is a green, strong, tall, grass. The grains on the head are rice like in upright facing clusters.

Ragwort is like a yellow flat daisy each flower having sixteen dark yellow petals. The toothed leaves are divided into narrow lobes tinged purple.

Horned poppy has yellow poppy like flowers and curved pointed leaves. It has very long curved thin seedpods.

Sea holly is vivid blue/grey with a thimble shaped arrangement of flowers. This is surrounded by a ring of blue/grey star shaped spiny leaves.

Red valerian is large plant with deep red/pink flowers very attractive to butterflies.

Sea lavender has white, pink or purple bell shaped flowers. The plant has woody stalks with clusters of flowers, blue or white, along one side.

VIKING BAY (A NATURAL HISTORY)

Please ensure that any stones lifted on the beech are replaced in the same position afterwards. Also, please put all rubbish in the bins especially plastics, which may not be broken down for hundreds of years or others that are eaten by fish etc. and enter the food chain.

MIKE PEARCE

To see other publications below by the author visit
snappysnappybooks.com

The really, really, really useful series
How to be a Successful Business Weed
How to Deal with Life's Snakes and Ladders
Know Your Students and Build Your Image
Pens for Pops
How to be a Successful Charity Shop
Make up-revealed
Ronnie's Sermon snippets
Wastefulness-Bone and Urine
Fertility Stones and Chocolate Eggs
Clingers, creepers and scramblers

Other books by Mike Pearce:
Pattern for Purpose- God's and Man's designs
Red Fred Cell and Friends
Human Termites eat London
Pigeons Splat London
Glass Anemones Tentacle-ize London
Tuppeny Hangover
I am Termite
The littlest Oyster
Bits and Bobs
The Shell Man
Cats at Christmas
Tails, Tales
Trust-Nothing but a Must
In a Dark, Dark Corner was the Holy Ghost
The Shell Lady

VIKING BAY (A NATURAL HISTORY)

Captain Grottbuster versus the Grey World
London's Nemesis (Trilogy of 3, 4 and 5 above)
Saved by Angels (Trilogy of 6, 8 and 14 above)
The World of Wax
Photosynthetic Women
Queen Rat on Deadman's Island
The Watcher on the Fal
The Rock Pool
The Little Shepherd Boy's Gift
The Living Fossils
Old Mother Nature Laughed and Laughed
Betty's Barcodes
Time Runs Dry (play)
Valentines Cards
The Scrofula Infirmary
The Cornish Urchin
My Therizinosaurus
Spider in the Tomb
The White Cockerel
The Red Church Doll
Butterfly Angels (compilation of previous books)
The Girl Under the Paeony Tree
Baby Feet
The Sparrows' Last Soul
I Herring Gull
Ball Rooms

ABOUT THE AUTHOR

Dr Mike Pearce is a scientist interested in behaviour. He also was a lecturer in human biology and health at a college in Canterbury, Kent.

www.ingramcontent.com/pod-product-compliance
Lightning Source LLC
Chambersburg PA
CBHW050030230526
45470CB00003B/1203